About this Series

Kayla's First Chickens is a series of four leveled readers designed for early readers. The series follows the adventures of Kayla as she (a) visits a farm, (b) raises her own hens, (c) accepts that it is normal to eat chickens, and (d) learns to prepare chicken for food. This series is part of a curriculum unit called *Food Origins.* Both the unit and this series of readers are aimed at helping children to develop better understandings of where our food originates.

About this Book

In *Kayla's First Chickens - Book One*, Kayla expresses interest in learning how to make chicken. In response, Kayla's father takes her to visit a farm so that she can learn more about chickens.

As a Level 1 Reader, *Kayla's First Chickens - Book One* uses simple words, offers picture clues, and introduces a limited number of new vocabulary terms. This text is a great learning resource for parents and teachers, and great fun for children!

Sankoré Institute

8631 Meridian Square Drive
Indianapolis, IN 46240

Visit our web site at www.sankoreinstitute.org

ISBN-13: 978-0-9789415-1-2

Printed in the United States of America

First Printing by Sankoré Institute: October 2014

To Princess Makayla Lillian Smith...

Education is what you know, not what's in the book.

~ African Proverb

SI Leveled Readers

1	Early Readers. These books are for beginning readers. They feature simple words, picture clues, and fewer vocabulary words. Here children continue their love of reading!
2	Developing Readers. These books are for shared reading. They feature advanced words, increased dialogue, and fewer vocabulary words. These books are both challenging and fun!
3	Independent Readers. These books are for newly independent readers. They feature longer sentences, short paragraphs, and an increasing number of vocabulary words. Children experience the joy and satisfaction of independence with these books!
4	Advanced Readers. These books are for advanced readers. They introduce children to chapters, they present challenging themes, and they offer great new vocabulary words. Children are off and running with these books!

Kayla's First Chickens

Book One

Jomo W. Mutegi

Kayla loves chicken. You can fry it. You can stew it. You can bake it.

Kayla loves chicken any way you make it.

Kayla asked her dad, “Can you teach me to make chicken?”

Her dad said, “Of course. But first, you need to learn how chickens are made.”

Kayla's dad took her to a farm. On the farm, Kayla saw a lot of chickens.

The farmer's daughter had a flock of ten chickens. She was learning how to raise them.

They had black and white striped feathers. They had yellow beaks. They had red wattles. They had red combs.

The farmer's son had five chickens. He raised them as a hobby.

They had brown feathers. They too had yellow beaks. They had red wattles. They had red combs.

The farmer's wife had two chickens. She kept them as pets. They had long, orange feathers on their necks. They had short orange feathers on their back capes. They had short orange feathers on their saddles.

They had black feathers on their breast. They had long green tail feathers. They had grey beaks. They had red combs. Kayla did not see a wattle. They laid pretty blue eggs.

The farmer had one flock of fifty chickens. He used these chickens to lay eggs.

They had white feathers. They had yellow beaks. They had red wattles. They had red combs.

- The farmer also had a flock of ninety chickens. He used these chickens for meat.

These chickens were big. They had black feathers. They had grey beaks. They had red wattles. They had red combs.

Kayla loved the chickens. Kayla watched the chickens.

She saw them scratch in the dirt to find food. They ate bugs. They ate worms. They ate small rocks.

Kayla played with the chickens. She saw them play with each other.

She saw them fight with each other. She saw them eat with each other.

Kayla ran with the chickens. They ran fast.

Sometimes they would fly. These chickens were so happy.

Kayla asked her dad, “May I raise some chickens at home?”

Her Dad said, “Of course.”

Kayla's dad bought two chicks from the farmer.

The farmer told Kayla, "Raise your chickens the natural way."

They all smiled and said, "Goodbye."

Discussion Topics

Below are a few good discussion topics. Discuss these topics with your child to encourage deeper thinking. It is not important to have "the right" answers. There are not always right answers. It *is* important to encourage children to think more about the reading.

1) Why did the farmer's family all have different types (or breeds) of chicken?

2) There are hundreds of breeds of chickens. Why do you think there are so many?

3) What other animal has many different breeds? Why are there so many?

New Words from Book 1

Kayla's First Chickens - Book One introduces early readers to a range of new words. Some of these words are sight words. Some are "science" words, in that they refer to parts of the chicken's anatomy. Others are words intended to extend the vocabulary of early readers.

It is important for parents and teachers to know that the way many of us were taught new vocabulary words in school is very ineffective. We were given a list of words each week. We took that list home and memorized the definitions and the spellings of those words. The following Friday we sat for our weekly vocabulary test. Very few new words are learned in this way. The best way to learn new words is to use them and to use them in a meaningful context.

A good strategy for helping children to learn new words is to encourage them to use the words. _One approach is to have discussions related to the text._ In these discussions, children should be encouraged to speak freely and should be encouraged to use new words. _A second approach is to engage children in activities around the text._ For example, you could have children write a letter to a family member describing how chickens eat. Or you could use colored pencils or acrylics to have children create a drawing or painting with chickens as the theme.

1. Back Capes	8. Natural
2. Beaks	9. Raise
3. Bought	10. Saddles
4. Breast	11. Scratch
5. Combs	12. Tail Feathers
6. Flock	13. Too
7. Hobby	14. Wattles

About the Author

Jomo W. Mutegi, Ph.D. is a mild-mannered professor by day and a relentless revolutionary by night. As a professor, Dr. Mutegi's area of expertise is science education. He studies the participation of African people in science, technology, engineering, and mathematics (STEM). He received a Bachelor's of Science degree in Chemistry and Biology from Gannon University; and he received a Doctorate in Science Education from Florida State University. He has published numerous manuscripts summarizing his research findings and theoretical arguments. He has taught thousands of prospective and practicing teachers, as well as K-12 students. He has helped to guide the research careers of several very accomplished STEM education researchers.

As a revolutionary, Dr. Mutegi is guided by the counsel of Dr. Amos Wilson who reminds us that, "The function of education is to secure the survival of a people." To this end he has committed himself to creating educational materials aimed at securing the survival the African people. These include curricular materials (e.g. lessons, modules, units, workshops, courses, activities, readers, and books) that can be used in both formal and non-formal settings. Most of these curricular materials focus primarily on STEM subject areas as they are relevant to African people.

Made in the USA
Monee, IL
05 November 2026

67375659R00021